Exploring Science

1

MANOJ PUBLICATIONS

Exploring Science-1

Publisher:
MANOJ PUBLICATIONS
761, Main Road, Burari, Delhi-110084 (INDIA)
Mobile : 09999476076, 09868112194,
08178823569, 08178854810
Email : info@manojpublications.com
for online shopping visit our websites:
www.sawanonlinebookstore.com

ISBN : 978-93-5579-328-7

Concept by:
Sanyam Gupta

Edited by:
Rohan Kumar

Preface...

The series entitled Exploring Science comprises a set of five books from classes I to V. Based on the NCERT syllabus, the series follows an activity-oriented thematic approach in which the child's learning is assessed in a comprehensive range of situations and environment both in and out of the classroom.

Each book of the series is divided into different topics. Each chapter starts with a synopsis of topics discussed under the heading Turning Points. Rack Your Brain is an explanation that makes the child feel aware about the topic and relates it to his daily life situations. The content of each chapter is developed in a simple and lucid way, and enriched with colourful illustrations. At the end of each chapter Summary of the Chapter is given to recapitulate the chapter.

The Experimental Work under the heading Get Hands-on Experience and Hands-on Learning is carried out in the form of Classroom Assignment, Home Assignment, Project Work and Classroom Presentation. This will help to link the textual knowledge with the practical experience.

We have made our sincere efforts to make this series error-free, refined and concise. The worthy suggestions for its further improvement are most welcome.

Contents...

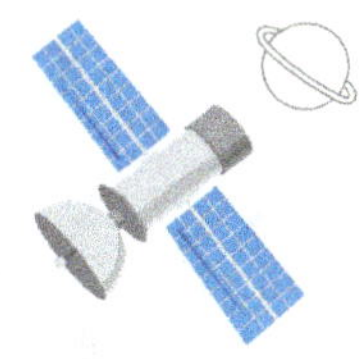

1 Our Body

We Will Learn

- Parts of the Body
- Our Sense Organs
- Growing up

Rack Your Brain

Do you ever wonder how we are able to talk and walk, read and write, jump and hop, sing and dance, and do so many other things? Can we do all the things with one part of the body? No, it's not possible for us. We do different activities with the different parts of the body.

Our body is like a wonderful machine. It is the combination of many parts. Each part of the body has a different name and function. Let us know more about our body-parts.

Parts of the Body

All our body-parts help us in different ways.

We walk, run and play with our legs. We draw, hold, write and push with our hands.

Walk

Run

Play

Write

Cut fruit

Eat

Board the school bus

Drink

Our Sense Organs

We have five sense organs : eyes, ears, nose, tongue and skin. They help us to see, hear, smell, taste and feel.

Eyes : Sight is an important sense. Our eyes help us to see. We have two eyes.

Ears : Hearing is also an important sense. Our ears help us to hear. We have two ears.

Nose : Our nose helps us to smell and breathe.

Tongue : Our tongue helps us to taste.

Skin : Our skin helps us to touch and feel.

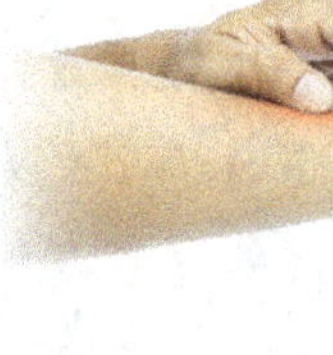

Brainy Point

Our eyes remain the same size even when we have grown up.

Growing Up

When we were born, we were small babies. A baby grows to become a child. The child grows into a young man and then into an old man. Young men or grown-up people are called adults.

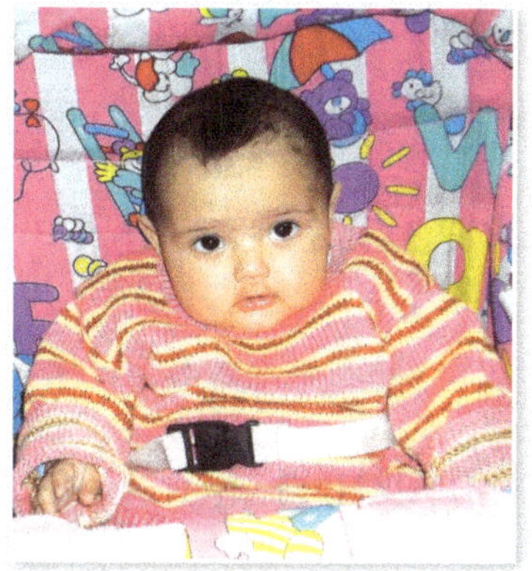
Baby

Boy

Man

Old man

In your family, you are a child and your parents and grandparents are adults.

Summary of the Chapter

- Our body is a combination of many parts.
- We do different things with the different parts of our body.
- Our sense organs help us to see, hear, smell, taste and feel.

A. Answer the following questions :

1. Name the five sense organs.

2. Write how our sense organs help us.

B. Fill in the blanks :

1. Our body is like a wonderful _______________ .
2. Our _______________ helps us in different ways.
3. We have five _______________ organs.
4. A baby grows to become a _______________ .
5. Our _______________ helps us to smell and breathe.

C. Tick (✓) the correct word :

1. Each part of our body does a **(different / same)** function.
2. We use our legs to **(eat / run)**.
3. Our tongue helps us to **(taste / touch)**.
4. We smell with our **(nose / ears)**.
5. Fingers are used for **(holding a pencil / kicking a football)**.

Get Hands-on Experience...

Take two paper bags. In the first bag, put three things which have a strong smell such as a sliced onion, an orange and a lemon-peel. In the second bag, put three things which have different tastes such as an apple, a biscuit and chips. Blindfold the children one by one and ask them to take out one item from one bag and identify it. Ask the other children to note the body-part he/she uses to find out what the item is.

Hands-on Learning...

Draw the pictures of sense organs in your scrapbook. Also, indicate the name of each sense organ.

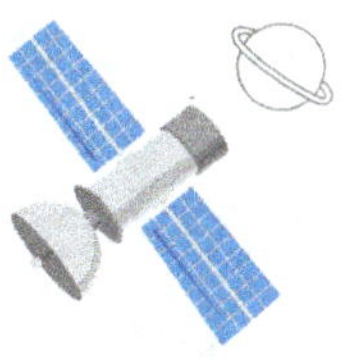

2 Food for Health

We Will Learn

- Need for Food
- Importance of Food
- Sources of Food

Rack Your Brain

We all need food to live; when we do not eat for long, we feel hungry. At that time, food is very essential for our body. It helps us to grow strong and do our tasks properly. Do you eat every type of food that your mom cooks at home or you like to eat pizzas, burgers, chips, etc.?

We should eat the food cooked by our mom as it includes the purity and her love which helps us to grow healthy and cheerful. Remember, a cheerful person always achieves success in every field. So, eat healthy and be healthy.

Let us first learn the things that we need to survive.

We need air, water and food to live and grow.

We need air to breathe. We breathe in and breathe out air all the time. All living things breathe in air. All plants and animals need air to breathe. We drink water when we are thirsty. We all need water to drink. Even plants need water to make their food.

Brainy Point

Every drop of water is important. You must not waste water. Take bucket-bath. Taking shower wastes a lot of water.

Need for Food

We need food to live and grow. Food keeps our body healthy and strong. If we do not eat for a long time, we feel hungry and weak. Therefore, we need different types of foods to keep ourselves healthy.

Importance of Food

1. Food helps us to stay healthy.

2. Food helps us to learn faster.

3. Food helps us to stay active.

4. Food gives us energy to play.

Brainy Point

The most common food in the world is cereals.

Sources of Food

Plants and animals are the main sources of food. We get vegetables, fruits, cooking oil, cereals and pulses from plants.

We get milk, eggs, honey and meat from animals. Milk is a complete food. We should drink milk daily. It makes our bones and teeth strong.

Summary of the Chapter

- Food helps our body to grow strong.
- Food gives us energy to work and play.
- We should eat all types of foods to keep ourselves healthy.

A. Answer the following questions :

1. Name three essential things for the survival of our life.

2. Why do we need food?

____

____

3. What are the uses of food?

____

____

4. Why should we drink milk daily?

____

____

5. Why do we need air?

____

____

6. When do we drink water?

____

____

B. Fill in the blanks :

1. We need air to ______________ .
2. We drink water when we are ______________ .
3. Plants and animals are the ______________ of food.
4. Food gives us ______________ to play.
5. Milk is a ______________ food.
6. Milk makes our ______________ and ______________ strong.

C. Give two examples for each of the following :

1. Food from Plants ____________ ____________
2. Food from Animals ____________ ____________

Get Hands-on Experience...

Go shopping with your mom and make a list of vegetables and fruits you see in the market.

Fruits	Vegetables

Hands-on Learning... 

Which are your favourite vegetables and fruits? Paste the pictures in the box given below.

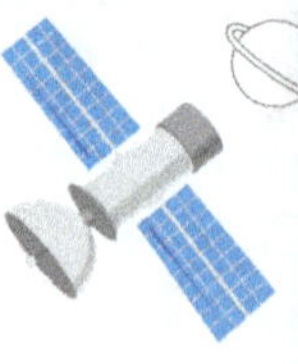

3 Keep Clean and Fit

We Will Learn

- Eating Healthy Food
- Suggestions for Good Health
- Good Habits and Keeping Clean

Rack Your Brain

Dear children! Let us read the story of a lazy boy, who didn't like to do any activity that was essential for keeping clean. He didn't like brushing, bathing, washing hands before and after every meal. He didn't like oiling and combing hair and also his nails were long and dirty. Can you imagine how he looked? How disappointed his parents were because of his bad habits! No one liked to talk with him. One day, the whole class boycotted him and then he realised the value of keeping clean; he promised to become a good child.

Eating Healthy Food

Green, fresh vegetables and fruits, cereals, pulses, milk, milk products, eggs and meat are called healthy foods because they help us to become strong and healthy.

We should not eat junk food as it makes us fall ill.

Burger

Pizza

Chips

Sauce

Coca Cola | Cake | Chocolate

Sweets | Noodles | Pasta

Suggestions for Good Health

- We should wash fruits and vegetables before eating them.
- We should never overeat.
- We should eat food at fixed times.
- Milk and milk products like curd, ghee, cheese, butter, etc. should be consumed every day. Milk makes our bones strong.
- We should exercise every day. It makes our body strong. Walking, running and swimming are good exercises. Playing games is also a good exercise. It builds our muscles.
- We should take proper rest to regain our energy. If we do not get enough rest, we feel sick.
- We should always drink clean water.

Brainy Point

We should drink 6-7 glasses of water every day.

- Windows should be kept open to let fresh air in.
- Play outdoor games and stay active.

Good Habits and Keeping Clean

- We should get up early and go to bed early too.
- We should brush our teeth twice a day.
- We should take a bath every day with soap and water.
- We should wear clean and well-ironed clothes.
- We should oil our hair before combing.
- We should wear polished shoes daily, polish them every night so that we may not get late for school.
- We should wash our hands before and after every meal.
- We should also not forget to wash our hands with soap after using the toilet.

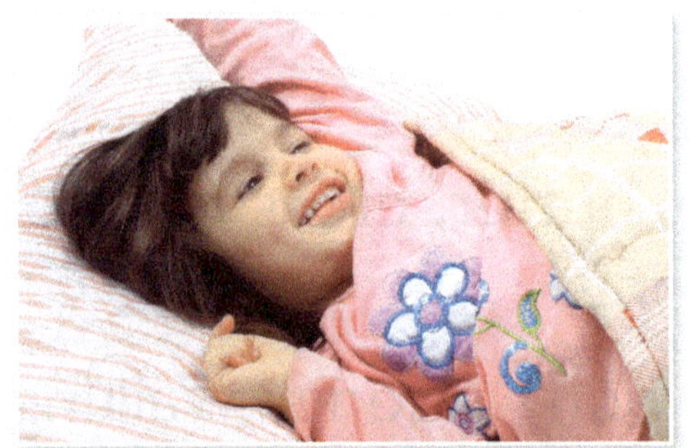

Brainy Point

Dirty socks smell bad and carry germs. Wear a fresh pair of socks every day.

Summary of the Chapter

- We should wash fruits and vegetables.
- Food gives us energy to work and play.
- We must keep our body clean.

A. Answer the following questions :

1. Name three healthy food-items.

____

____

2. Name three junk food-items.

____

____

3. Why should we take proper rest?

____

____

4. Write any two good habits that are essential for keeping clean.

____

____

B. Fill in the blanks :

1. Milk makes our ____________ strong.
2. ____________ food makes us fall ill.
3. ____________ is a milk product.
4. ____________ , ____________ and ____________ are good exercises.

C. Match the following :

Junk Food

Healthy Food

D. Write (T) for a true statement and (F) for a false one :

1. We should drink plenty of water.
2. We should polish our shoes early in the morning.
3. Junk food makes us feel healthy.
4. Only plant food is essential for good health.

Get Hands-on Experience...

Don't throw away water used for boiling vegetables or washing food-items. Use it to water your plants.

Hands-on Learning...

Take a thermocol plate. Paste a red glaze paper on it. Draw an apple on it and cut it out neatly. Paste the apple to a straw. Your apple puppet is ready.

4 Safety Rules

We Will Learn

- Safety
- Safety Rules

Rack Your Brain

Dear children, sometimes when we are happy and excited, we start doing mischievous activities and also become careless. We forget that we can get hurt anywhere. We should remain careful at home, at school or on the road. We can be safe only when we are careful.

Let us study about being safe and safety rules in this chapter.

Safety

Safety means to remain safe from any harm. We can remain safe if we are careful all the time.

To be safe, we should follow some safety rules at every place and every moment.

Safety Rules

Safety rules help to keep us safe. Learn the following safety rules.

While Walking on the Road

- Do not run on the road.
- Always keep to your left.
- Always walk on the footpath.

While Crossing the Road

- Cross only when the signal light is green.
- If there is no signal, look to your left, then to your right and then to your left again and cross the road if it is clear.
- If there is the zebra crossing then we should cross the road only at the zebra crossing.

While Boarding a Bus

- Always stand in a queue while waiting for the bus.
- Wait for your turn patiently.
- Do not push anybody while getting into or out of the bus.
- Do not lean out of the window of the moving bus.

- Do not throw things out of a moving bus. Someone may get hurt.
- Do not shout or move inside a moving bus.
- Do not get into or off a moving bus. You may slip and hurt yourself.

Brainy Point

When we shout or move in a bus or car, it distracts the driver and might cause an accident.

While Playing

- Do not play games that may harm you.
- Do not play on the road. Always play in a park or playground.
- Do not pluck the flowers or leaves of plants. There may be thorns with flowers.

While Swimming

- Never go swimming alone. Always swim with an adult around.

Safety Rules at School

- Do not sharpen your pencil with a knife or cutter.
- Do not jump on chairs and desks in your classroom. Do not run inside your classroom.
- Do not run up and down stairs. You may fall and get hurt.
- Do not push or pull your friends even for fun.
- If you or any of your classmates gets hurt, inform your teacher immediately.

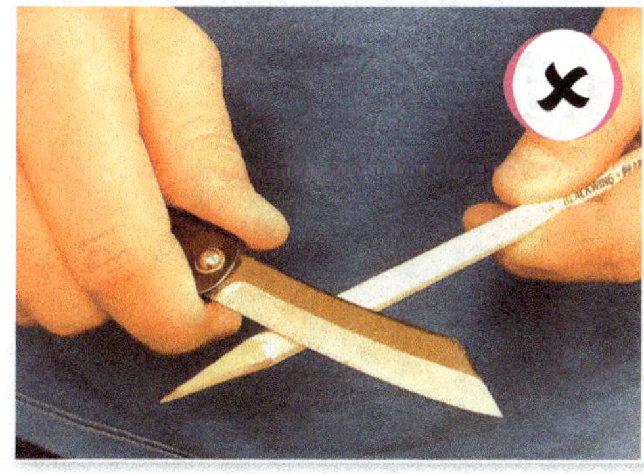

Safety Rules at Home

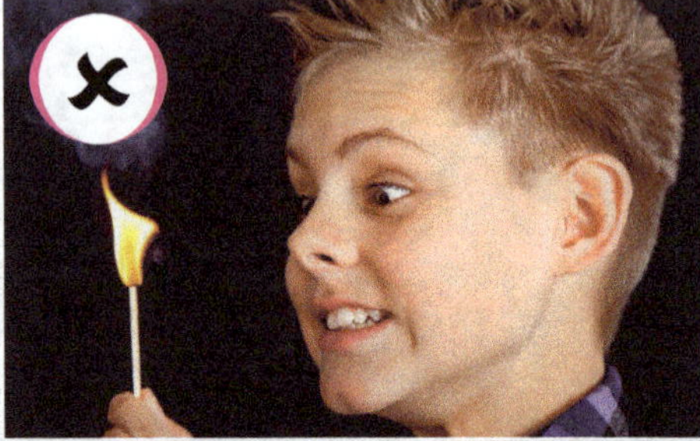

- Do not play with sharp things such as a knife, scissors or a needle. You may hurt yourself.
- Do not touch electric switches or wires with wet hands.
- Do not play with matchsticks. Stay away from fire.
- Keep your toys and other things at proper places. Do not leave them on the floor as anyone may slip over them.
- Do not climb on shelves. You might fall and get hurt.

Brainy Point

We should not touch electric switches with wet hands as we might get electric shocks.

Summary of the Chapter

- We must follow safety rules to be safe.
- Safety rules should be followed everywhere whether we are on the road, in the bus, at home or playing in the park.
- We should cross the road when the signal light is green.

A. Answer the following questions :

1. What do you mean by safety?

2. Write two points that help us to remain safe at home.

____

3. How should we cross the road?

____

4. Write two safety rules that you should follow at school.

____

B. Fill in the blanks with appropriate words:

1. ____________ help to keep us safe.
2. We should always walk on the ____________ side.
3. Always play in a ____________ or ____________ .
4. Do not leave your toys on the ____________ .
5. Always stand in a ____________ while waiting for a bus.

C. Cross out (×) the wrong words :

1. We (must / must not) run on stairs.
2. Do not get into or off a (standing / moving) bus.
3. Electric switches should not be touched with (wet /dry) hands.
4. We should cross the road at the (zebra / giraffe) crossing.
5. (Use / Do not use) a knife to sharpen your pencil.

Get Hands-on Experience...

Make a list of fire stations, police booths and hospitals in your locality. Also, mention their phone numbers.

Hands-on Learning...

Make a chart on safety rules at school. Display the chart on the display board of your class.

5 Animals Around Us

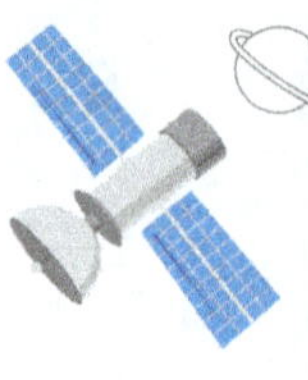

We Will Learn

- Animals in our Home
- Animals in Forest
- Animals that Fly in the Air
- Insects
- Animals that Live in Water
- Animals that Live Both on Land and in Water

Rack Your Brain

We see many animals around us. They all are different in sizes and nature but these three things are common in all animals– they all breathe, they all need food to grow, they all produce their young ones. But all animals have their own lifestyle.

There are different kinds of animals in the world. They live at different places and eat different types of foods.

Let us read about the living places of animals.

Animals in Our Home

The animals that live in our homes are called pet animals.

Examples:

Cat

Dog

Rabbit

Parrot

The animals that help us in our work are called domestic animals.

Examples:

Cow
(gives milk)

Horse
(carries heavy loads)

Sheep
(gives us wool)

Ox
(ploughs the field)

Animals in Forest

The animals that live in a forest are called wild animals. Wild animals are of two types: Plant-eating animals and flesh-eating animals.

The animals that eat only plants are called plant-eating animals.

Examples:

Elephant Zebra Deer Giraffe

The animals that eat only flesh are called flesh-eating animals.

Examples:

Lion Jackal Wolf Fox

The bear and the crow eat both plants and flesh.

Examples:

Bear

Crow

Animals that Fly in the Air

Birds fly in the air. They have wings that help them to fly. They have two legs which help them to walk. They do not have teeth; they eat with their beaks.

Parrot

Eagle

Ostrich

Pigeon

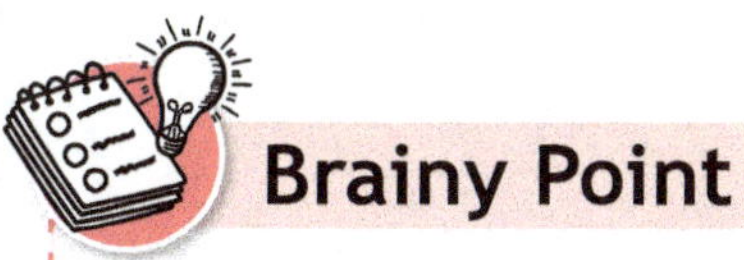

Brainy Point

The ostrich, the penguin and the emu are flightless birds.

Insects

Insects are very small animals. They have six legs. Most of the insects have wings. But the insects like ants, bedbugs and beetles do not have wings.

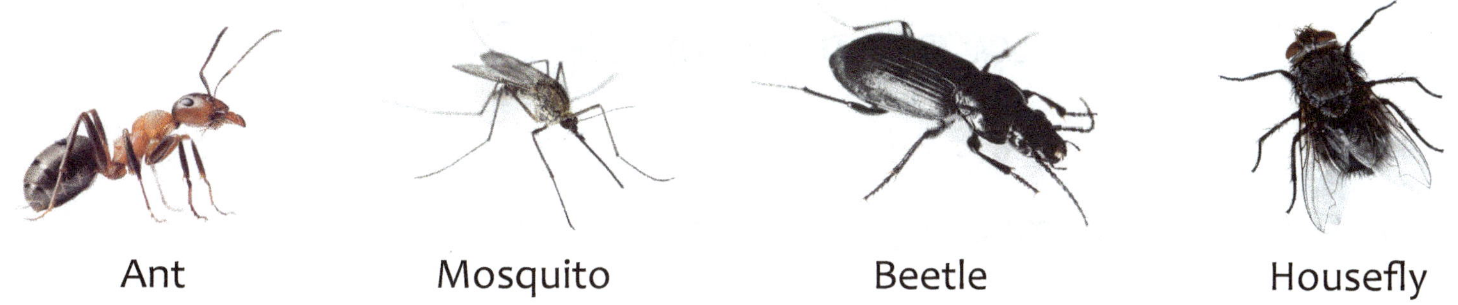

Ant Mosquito Beetle Housefly

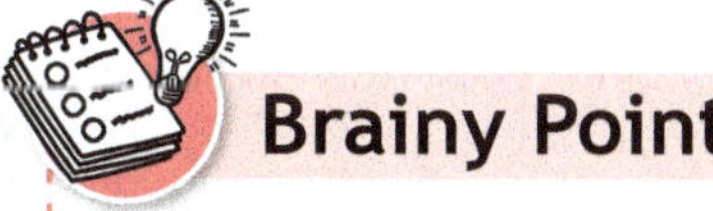

Brainy Point

Spiders are not insects as they have eight legs.

Animals that Live in Water

The animals that live in water are called water animals.

Whale Octopus Shark

Animals that Live both on Land and in Water

Some animals can live both on land and in water.

Crab Turtle

Frog

Crocodile

Summary of the Chapter

- The animals that live in our homes are called pet animals.
- Wild animals live in a forest.
- The animals that live in water are called water animals.
- Some animals can live both on land and in water.

A. Answer the following questions :

1. What do you mean by plant-eating animals?

2. What do you mean by flesh-eating animals?

3. How many legs do insects have?

4. Write three lines about birds.

B. Fill in the blanks with right words:

1. The dog is a ______________ animal.
2. An ______________ is a domestic animal.
3. ______________ animals are of two types.
4. Birds ______________ in the air.
5. An ______________ does not have wings.

C. Match the following :

	Column A		Column B
1.	Bird that can fly	a.	Beetle
2.	Bird that cannot fly	b.	Moth
3.	Animal that lives on land	c.	Emu
4.	Animal that lives in water	d.	Elephant
5.	Insect that can fly	e.	Pigeon
6.	Insect that cannot fly	f.	Hippopotamus

Get Hands-on Experience...

Visit the park near your home and observe the different types of insects. Do you find any similarity among them?

Hands-on Learning...

Paste the pictures of different animals on a chart paper and also write their names beautifully.

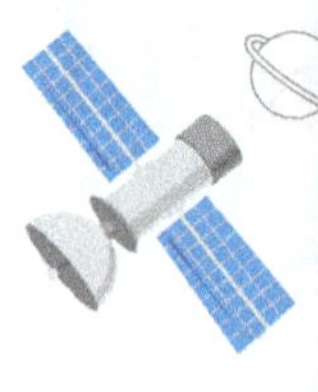

6 Food and Homes of Animals

We Will Learn

- Food of Animals
- Homes of Wild Animals
- Homes of Pet Animals
- Homes of Domestic Animals
- Homes of Animals Living in the Open

Rack Your Brain

Dear children! There are different kinds of animals in the world. They all need food to live and grow. They need shelter as well. We see many animals in the zoo; but we do not see them around us.

Let us read about all these animal shelters and food.

Food of Animals

Different animals eat different kinds of foods. Some animals eat only plants.

Cow

Deer

Elephant

Some animals eat grains.

Squirrel

Pigeon

Hen

Some animals eat insects.

Spider

Lizard

Frog

Some animals eat the flesh of other animals.

Lion

Wolf

Jackal

Some animals eat both plants and the flesh of other animals.

Bear

Crow

Homes of Wild Animals

Just like man, an animal also needs a home to live in. Animals need homes to be safe from heat, cold, rain and enemies.

A lion lives in a den.

An elephant lives in a forest.

A snake lives in a hole.

Homes of Pet Animals

We make special homes for our pets.

Parrots are kept in cages.

Dogs are kept in kennels.

Hens are kept in coops.

Homes of Domestic Animals

Cows are kept in sheds.

Horses are kept in stables.

Sheep are kept in pens.

Homes of Animals Living in The Open

Some animals make their own homes.

Bees make a bee-hive.

A spider spins a web.

Birds make their nests.

Butterflies live in a garden.

Ants make an anthill.

A woodpecker makes a hole in the bark of a tree.

Summary of the Chapter

- Animals eat food for energy, growth and strength.
- Some animals eat plants only, some eat the flesh of other animals and some eat both plants and animals.
- Animals need homes to live in.

A. Answer the following questions :

1. Which animal makes a web?

2. Where does a woodpecker live?

____

3. What is the home of hens called?

____

B. Tick (✓) the correct answer :

1. Which of the following animals eats grains?

   a. Lion ☐ b. Pigeon ☐ c. Bear ☐

2. Which animal lives in a forest?

   a. Elephant ☐ b. Horse ☐ c. Both a & b ☐

3. Which of the following animals is kept in a pen?

   a. Cow ☐ b. Hen ☐ c. Sheep ☐

C. Fill in the blanks :

1. Frogs and lizards eat _______________.
2. Hens eat _______________.
3. We keep parrots in _______________.
4. A snake lives in a _______________.
5. Some animals make their own _______________.

Get Hands-on Experience...

Talk about your pet if you have one at home. If you do not have a pet, would you like to keep one? Which animal do you like for a pet? Why? What would you name it?

Hands-on Learning...

Collect the pictures of some animals with their homes. Paste them in your scrapbook.

7 Our Surroundings

We Will Learn

- Our Surroundings
- Keep the Surroundings Clean
- Make the Surroundings Green

Rack Your Brain

John and Riya were going to school with their father. John was peeping out of the car. He saw many trees, animals and buildings on the way. He asked his father, "Dad, we can see many things around us. What are these things called?" "These things are called surroundings," replied his father.

Let us read about our surroundings in this chapter.

We see many things around us, such as people, places, vehicles, plants, trees, animals, etc. All these things make our surroundings.

Our Surroundings

The places and the things that are around us make our surroundings.

In a classroom, we have classmates, chairs, tables, books, pencils, blackboard, chalk, bags, etc. in our surroundings.

In a playground, we can see trees, plants, flowers, our friends, swings, etc. in our surroundings.

On a road, we see vehicles like cars, bikes and trucks in our surroundings.

Keep the Surrounding Clean

We should keep our surroundings clean and tidy.

We can stay healthy only in clean surroundings.

These are the ways to keep our surroundings clean–

1. Keep all the things at proper places in your room.
2. Throw all the waste things in the dustbin.
3. Clean all the things of your room with a duster.

4. Don't throw garbage out of your house.
5. Grow plants near your house.

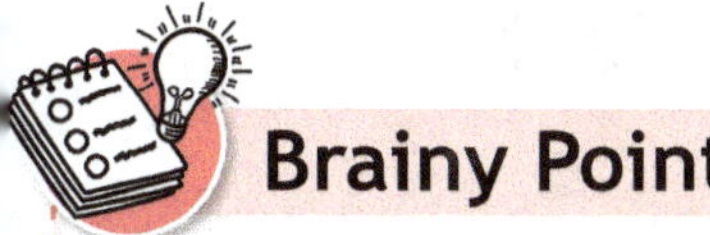

Brainy Point

Chandigarh is the cleanest and greenest city in India.

Make the Surroundings Green

Plants provide fresh air to breathe. They also provide cool wind and shade during summer.

If we don't have the space for trees, we can keep big pots with beautiful and useful plants outside our house.

We should take proper care of these plants as they make our surroundings clean and green.

Summary of the Chapter

- The things around us form a part of our surroundings.
- We should keep our surroundings clean and tidy.

Let's Comprehend

A. Answer the following questions :

1. What do you mean by surroundings ?

2. Why should we keep our surroundings clean?

____

3. Write any two ways of keeping our surroundings clean.

____

4. What do plants provide us?

____

5. Why should we take proper care of plants and trees?

____

B. Fill in the blanks :

1. We see many __________ around us.
2. Plants and trees are also the part of our __________.
3. We see a blackboard in a __________.
4. We should throw waste things in a __________.
5. We should keep our surroundings __________.

C. Match the following :

	Column A		Column B
1.	Notebook	a.	Corner Table
2.	Tumblers and Plates	b.	Cupboard
3.	Table Lamp	c.	Bag
4.	Toys	d.	Kitchen
5.	Clothes	e.	Rack

D. Name two things that you see:

1. in a park ______ ______
2. on the road ______ ______
3. on your study table ______ ______
4. in the classroom ______ ______

Get Hands-on Experience...

Visit your neighbourhood with your parents. Make a list of the things that you saw in your neighbourhood as a part of your surroundings.

Hands-on Learning...

Make your own tree. Gather some dry leaves from a park. Draw a tree-trunk and branches on a A-4 size sheet with a brown crayon. Glue the dry leaves onto the branches. Your tree is ready.

8 The World of Plants

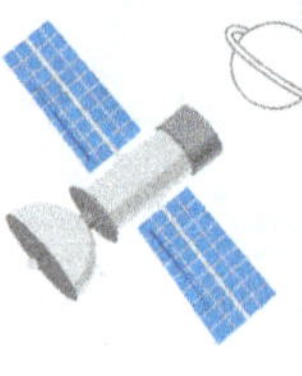

We Will Learn

- Herbs
- Shrubs
- Trees
- Climbers
- Creepers

Rack Your Brain

Dear Children! Look around your house. You will see many plants. These are of different shapes and sizes. There may be very big plants, like mango tree, small plants like a rose plant or very small plants like the grass.

Let us read about them.

Our surroundings look beautiful when there are plants all around us. Let us learn more about plants according to their sizes.

Herbs

Very small and weak plants are called herbs. These plants have soft stems. These plants live only for three to four months.

Grass

Coriander

Spinach

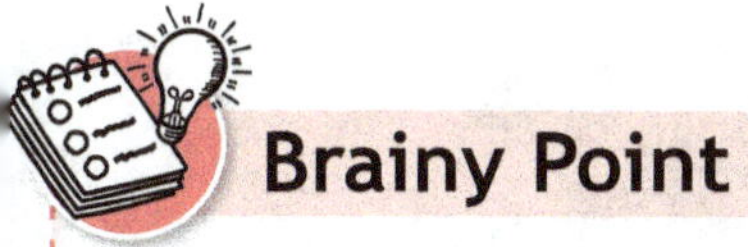

Brainy Point

Herbs like tulsi are used as medicines.

Shrubs

Small plants with strong stems are called shrubs. These are short and bushy plants. Most shrubs live for only a few years.

Rose

Croton

Hibiscus

Trees

Very big, tall and strong plants are called trees. They have thick and woody stems.

Peepal

Coconut Tree

Banyan Tree

Climbers

The plants which cannot stand on their own, are called climbers. Climbers need some support to grow as they have weak and thin stems.

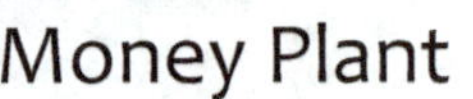

Money Plant

Grapevine

Pea Plant

Creepers

Some weak plants grow along the ground. They have very weak and thin stems. These are called creepers.

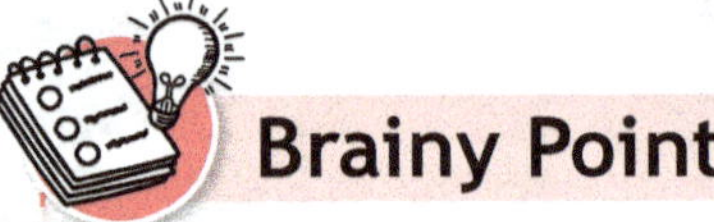

Brainy Point

Money plants grow well in shade.

Pumpkin

Watermelon

Bitter Gourd

Summary of the Chapter

- There are different kinds of plants.
- Big plants are called trees.
- The smaller bushy plants are called shrubs.
- The smaller plants are called herbs.

Let's Comprehend

A. Answer the following questions :

1. What are trees?

___

2. What are shrubs?

___

3. What are herbs?

___

4. What is the difference between a creeper and a climber?

___

B. Fill in the blanks :

1. Herbs have ________ stems.
2. ________ are short and bushy plants.
3. Trees have ________ and ________ stems.
4. ________ are used for making medicines.
5. ________ grow along the ground.

C. Give two examples for each of the following :

1. Trees ________ ________
2. Creepers ________ ________
3. Herbs ________ ________
4. Shrubs ________ ________
5. Climbers ________ ________

D. Match the following :

	Column A		Column B
1.	Hibiscus	a.	Tree
2.	Coriander	b.	Shrub
3.	Banyan	c.	Climber
4.	Watermelon	d.	Herb
5.	Pea Plant	e.	Creeper

Get Hands-on Experience...

Collect the pictures of some plants and paste them in your scrapbook. Classify them as shrubs, herbs and trees.

Hands-on Learning...

Identify the following plants and write their names:

1. It is a shrub. It bears the flowers of many colours. Children like its red flowers very much. But don't try to pluck it. You may get hurt.

   ____

2. It is a big tree with many hanging roots. Many birds make their nests on its branches. It gives shade to many passers-by during summer.

   ____

9 Parts of a Plant

We Will Learn

- Root
- Stem
- Leaf
- Flower
- Fruit
- Seed
- Growth of a Plant

Rack Your Brain

Just like a human body, plants also have different parts. These parts work together to help plants grow and stay alive. Each part performs a different kind of work.

Let us know the various parts of a plant.

Look at the given picture and observe the plant carefully.

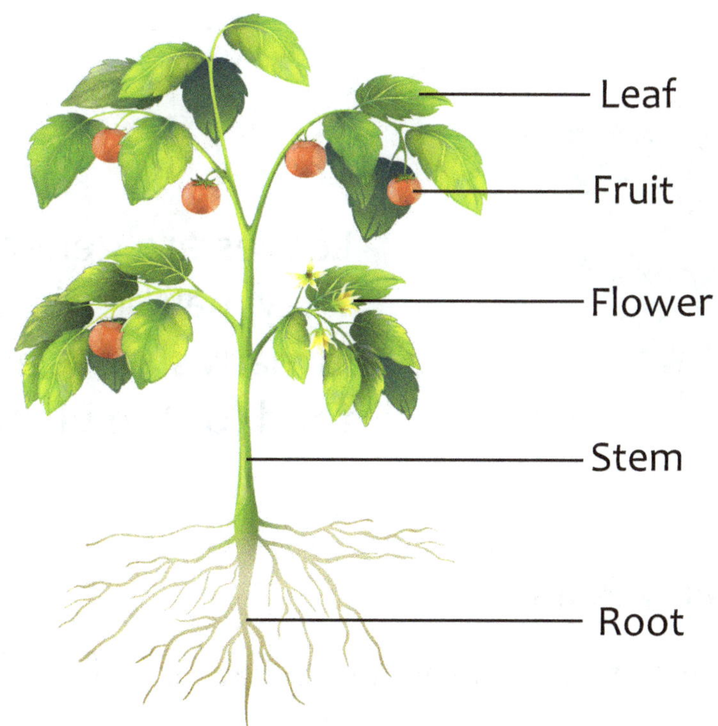

Root

The root is the underground part of the plant. It fixes the plant into the soil and allows it to grow straight. It is generally brown in colour.

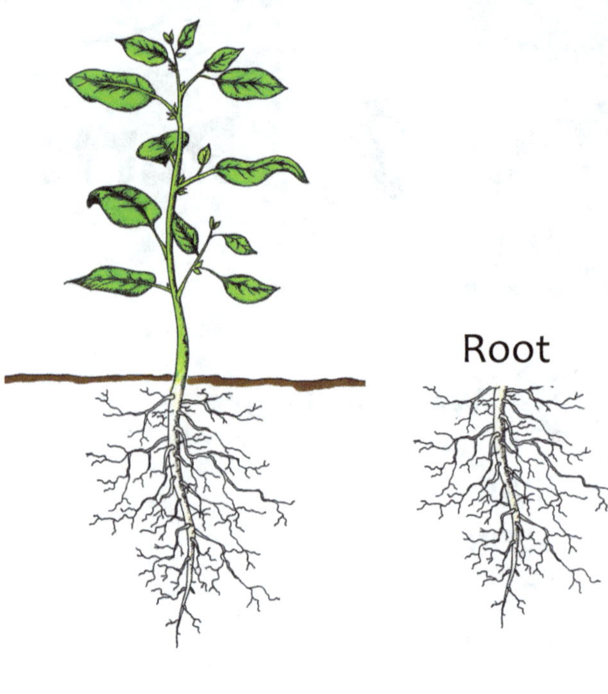

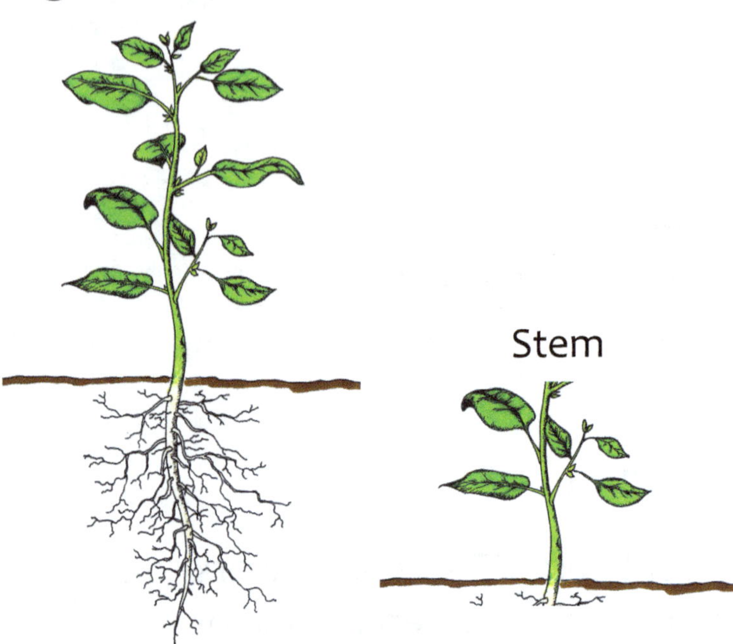

Stem

The stem is the main part of the plant. It grows above the ground.

The stem has branches with leaves, flowers, fruits and buds on it.

Brainy Point

Some plants have underground stems as well.

Leaf

Leaves are generally flat and green. They may be big or small. They come in many shapes and sizes. They make food for the plant.

Brainy Point

Leaves are known as the 'Food Factory of the Plant'.

Flower

A bud grows into the flower. Flowers are the most beautiful part of the plant. They usually have a nice smell.

Flower

Fruits

Fruit

Fruits grow from flowers. They are eaten as food. Fruits are found in different shapes and sizes.

Seed

Seeds are found inside fruits. A fruit may have only one seed whereas some fruits have many seeds inside them. A seed has a baby plant inside it.

Seeds

Growth of a Plant

Most plants grow from seeds. When we sow a seed into the soil, it needs three things to grow into a plant. These are :

- Air
- Water
- Sunlight

The air helps a plant to breathe.

Water helps to let it grow. Water comes from the rain.

Sunlight keeps it warm. A seed gets sunlight from the sun.

When a seed gets all these things, it begins to grow into a plant.

Summary of the Chapter

- All plants, big or small, have roots, stems, branches, leaves, flowers and fruits.
- The root grows deep into the soil.
- Leaves have tiny holes through which they breathe.
- Flowers change into fruits.
- Seeds grow into new plants.

A. Answer the following questions :

1. Which part allows a plant to grow straight?

____

____

2. Which part grows into a fruit?

____

3. What are the three things a seed needs to grow into a plant ?

____

B. Fill in the blanks :

1. The _______________ fixes the plant into the soil.
2. The stem grows _______________ the ground.
3. Leaves make _______________ for the plant.
4. _______________ have usually a nice smell.
5. A fruit may have only one _______________ .

C. Tick (✓) the correct answer :

1. It grows into a flower:

   a. Seed ☐ b. Bud ☐ c. Leaf ☐

2. This is called the 'Food Factory of the Plant':

   a. Seed ☐ b. Bud ☐ c. Leaf ☐

3. This part is eaten as food:

   a. Stem ☐ b. Leaf ☐ c. Fruit ☐

4. It helps plants to breathe:

   a. Water ☐ b. Air ☐ c. Sunlight ☐

5. Leaves, flowers and fruits grow on it:

   a. Stem ☐ b. Root ☐ c. Bud ☐

Get Hands-on Experience...

Leaves make food for the plant. Ask your teacher the things that leaves need to make food.

Hands-on Learning...

Take a small pot. Fill it up with garden soil. Sow any seed in it. Keep it out in the sun and water it. Observe it carefully. After two or three days, you will see a seedling growing out of it. Take care of it just as your mom takes care of you.

10 Plants As Food

We Will Learn

- Fruits
- Dried Fruits
- Vegetables
- Cereals and Pulses
- Spices
- Edible Oil
- Coffee, Tea and Sugar

Rack Your Brain

Dear children! We all need food to stay alive. Do you know where this food comes from? Most of the food, we eat, comes from plants. We get all this food in different forms.

Let us read about the different parts of a plant as food.

Fruits

We get many fruits from plants. These are eaten raw. They are mostly soft, fleshy and tasty. They are also good for our health.

Mango Apple Orange Grapes

Dried Fruits

The fruits that are eaten when they are hard and dry, are called dried fruits. They give strength to our body.

Almond Cashewnut Pistachio Walnut

Vegetables

Plants give us vegetables. We cook vegetables to make them edible but some vegetables are eaten raw too.

Tomato Cauliflower Potato Carrot

Eating carrots is very healthy for eyes.

Vegetables are roots, stems, leaves and flowers of plants.

Cereals and Pulses

Cereals and pulses are also called foodgrains. These are the seeds of the plant that we eat. Some foodgrains are cooked as pulses and rice. Some are used as flour such as wheat and maize.

Wheat Rice Maize Gram

Spices

Spices improve the taste and flavour of the food, so they are very important for food.

Turmeric Chilli Cumin seeds Coriander

Brainy Point

Turmeric can be added into a drink to treat cold and stomach complaints.

Edible Oil

Edible oil is used for cooking food. We get different edible oils from different plants.

Mustard → Mustard Oil Groundnut → Groundnut Oil

Coffee, Tea and Sugar

Coffee, tea, sugar, cocoa and chocolate powder are some other eatables which we get from plants.

Tea leaves

Tea

Chocolate powder

Sugarcane/Sugar

Thus plants are very useful to us.

Summary of the Chapter

- We get many fruits from plants.
- Dried fruits give strength to our body.
- Cereals and pulses are also called foodgrains.
- Edible oil is used for cooking food.
- Plants are very useful to us.

A. Answer the following questions :

1. From where do we get most of our food?

   ____

2. What are dried fruits?

   ____

3. What are foodgrains?

   ____

4. Why are spices important for food?

   ____

5. Which grains are used for making flour?

____

B. Fill in the blanks with the suitable words:

1. __________ are our green friends.
2. Fruits are eaten __________ .
3. __________ is an example of the root that we eat.
4. __________ and __________ are used as flour.
5. Sugar is made from __________ .

C. Give two examples for each of the following :

1. Cereals __________ __________
2. Pulses __________ __________
3. Vegetables __________ __________
4. Fruits __________ __________
5. Edible oils __________ __________
6. Dried fruits __________ __________

D. Tick (✓) the correct answer :

1. We get oil from:
   a. sunflower ☐
   b. cauliflower ☐
   c. banana flower ☐
2. We get coffee from:
   a. plants ☐ b. factories ☐ c. animals ☐

3. Turmeric, coriander and chilli are some examples of:

a. oilseeds ☐ b. pulses ☐ c. spices ☐

4. We get cereals and pulses from:

a. animals ☐ b. plants ☐ c. fields ☐

Get Hands-on Experience...

Make salad (fruits and vegetables) in the class and enjoy the party.

Hands-on Learning...

A. Draw any two healthy food-items:

B. Make a peacock of pulses:

Make an outline of a peacock on a drawing sheet. Spread glue on it. Now, paste the pulses of different colours, like arhar and malka masoor. Use urad dal to make the peacock's eyes. Let them dry. The colourful peacock you have drawn is now ready to dance.

11 Air and Water

We Will Learn

- Air Around Us
- Features of Air
- Uses of Air
- We Need Water
- Uses of Water
- From Where do We Get Drinking Water?

Rack Your Brain

Salma was sitting in a closed car. After a while, she started feeling sick; she got out of the car to breathe some fresh air. She drank some water. She felt fine again. Why do we need air and water?

Let us read in the chapter.

Air Around Us

Air is all around us, but we cannot see it. We can only feel the air when it moves.

Moving air is called wind. Wind can move things.

Features of Air

- Air occupies space. A balloon inflates when we blow air into it.
- Air has weight. An inflated balloon is heavier than a deflated one.

Uses of Air

1. All living things need air to breathe and survive.

2. Air is helpful in sailing boats.

3. Air is helpful in the movement of windmills.

4. Air is needed for burning.

5. Air is needed for drying clothes.

6. Air is filled in the tyres of vehicles.

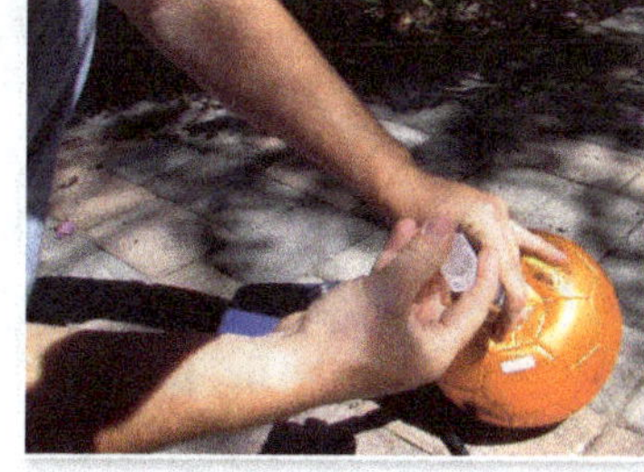

7. Air is used for inflating footballs and balloons.

8. Air helps to fly kites.

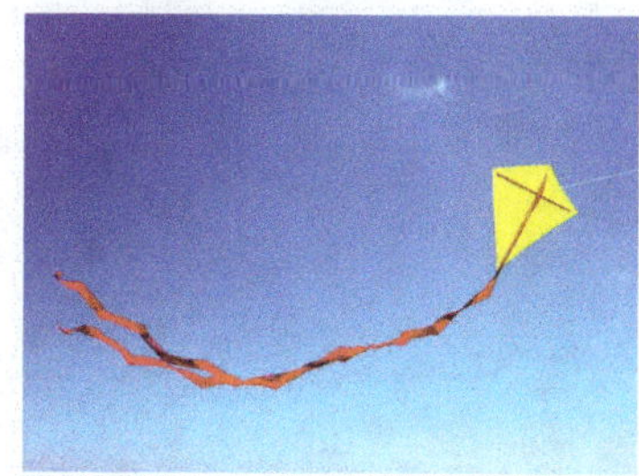

We Need Water

All living things need water to stay alive. We should drink plenty of water to grow healthy.

Uses of Water

Water is used by all living things–animals, plants and human beings.

People use water for various purposes. They use it to drink, cook food, wash clothes and utensils and to clean body. It is also used for putting out fire.

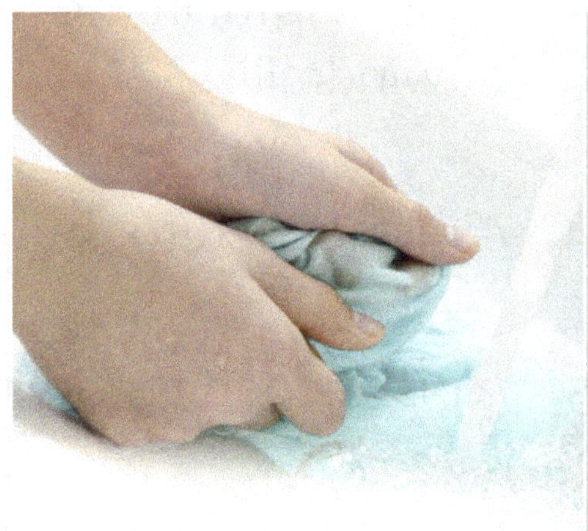

From Where do We Get Drinking Water ?

Rain is the main source of water. Rainwater fills up rivers, ponds and lakes. Most rivers join the sea. The water from rivers, ponds and lakes is collected, cleaned well and it finally reaches our homes through taps.

If the water that we get is not clean, we must boil the water and then use it.

Brainy Point

A person can live for a month without food, but only for a week without water.

Summary of the Chapter

- All living things need air.
- Air is present everywhere on the Earth.
- We need water to drink, to take a bath, to cook and to water plants.

A. Answer the following questions :

1. What do you mean by wind?

____

2. What are the uses of air ? Write any two.

____

3. Write the uses of water in our daily life.

____

4. From where do we get water ?

____

B. Fill in the blanks with the suitable words:

1. We ______________ see air.
2. We all need air to ______________.
3. Air occupies ______________.
4. ______________ is the main source of water.
5. Most ______________ join the sea.

C. Write the missing letters to complete the words :

P				S

R		V			S

L		K		S

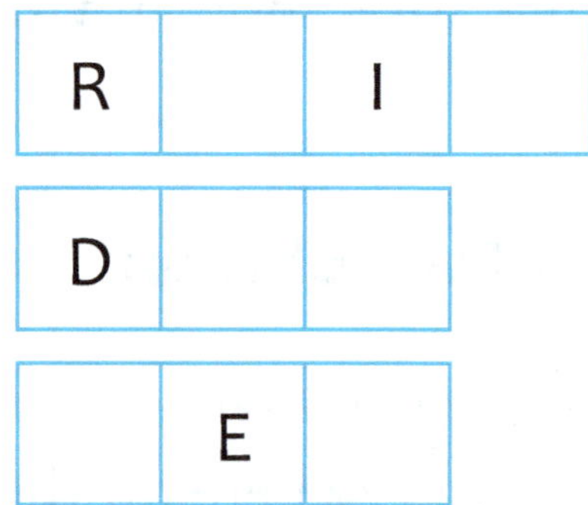

Get Hands-on Experience...

Do an activity to show that air takes up space.

Take a balloon and blow air into it. It increases in size. Why? Because the air that you blow into the balloon gives shape to it. This shows that air occupies space.

Hands-on Learning... 

Make a kite with light paper taking the help of your parents. Fly it in a park and observe how wind affects the direction of the kite.

12 The Sun, the Moon and Stars

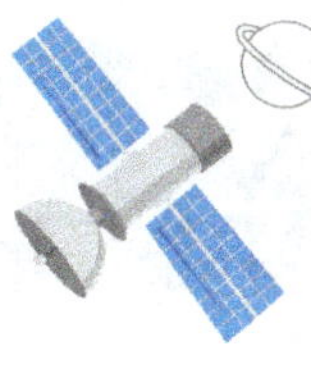

We Will Learn

- The Sun
- The Moon
- The Stars

Rack Your Brain

Monal was going home from school in the afternoon. The Sun was shining and it was very hot. He was getting irritated and thought there should be no Sun in the sky. Then he asked his teacher why we need the Sun. His teacher explained that the Sun is essential to all of us. Do you know why the Sun should be there?

What does the Sun give us? How does it help all living beings? When does it rise and set? What happens when the Sun sets? Let's find out all the answers in this chapter.

The Sun

The Sun is round and bright.

It is like a big hot ball of fire.

It looks yellow in colour.

It gives us heat and light.

The heat of the Sun gives life to all plants and animals.

Every morning, the Sun rises in the east. And with the rise of the Sun, a new day begins for all. There is light all around during the day.

Every evening the Sun sets in the west. Sunset brings night. There is darkness all around during the night.

The Moon

We see the Moon in the sky at night. It is smaller than the Sun. The Moon is round like a ball. It changes its shape every night. Sometimes it looks round. Sometimes it is C-shaped. Sometimes it disappears completely.

The Stars

We see a lot of twinkling stars in the sky at night. The stars are very big balls of fire. They are very far away from us; that is why they appear so tiny. The Pole Star is a star which is static in the northern sky. It is the brightest star in the sky. Some groups of stars make some patterns in the sky. For example, the shapes of a Polar Bear and a Hunter.

Brainy Point

Stars do not actually twinkle. They appear to twinkle due to turbulences in the Earth's atmosphere.

Summary of the Chapter

- The Sun is a big ball of fire. It gives us heat and light .
- The Moon is a bright silvery ball-like object that shines at night.
- A star is a huge sphere of very hot, glowing gases.

A. Answer the following questions :

1. How is the Sun's heat useful to all plants and animals?

2. What are the different shapes of the Moon?

3. Why do the stars appear so tiny to us?

B. Write the words correctly in the Sun Box and the Moon Box :

school	dinner	sleep	play	stars	sunlight
dark	lunch	dream	breakfast	kite	owl

Sun Box	Moon Box

C. Fill in the blanks :

1. The Sun gives us ________________ and ________________ .
2. The Sun sets in the ________________ .
3. The Moon is ________________ than the Sun.
4. The stars appear very ________________ to us.

Get Hands-on Experience...

Use your imagination and observation and create a night sky using a black sheet and silver paper.

Hands-on Learning... 

Draw the phases of the Moon in the given circles:

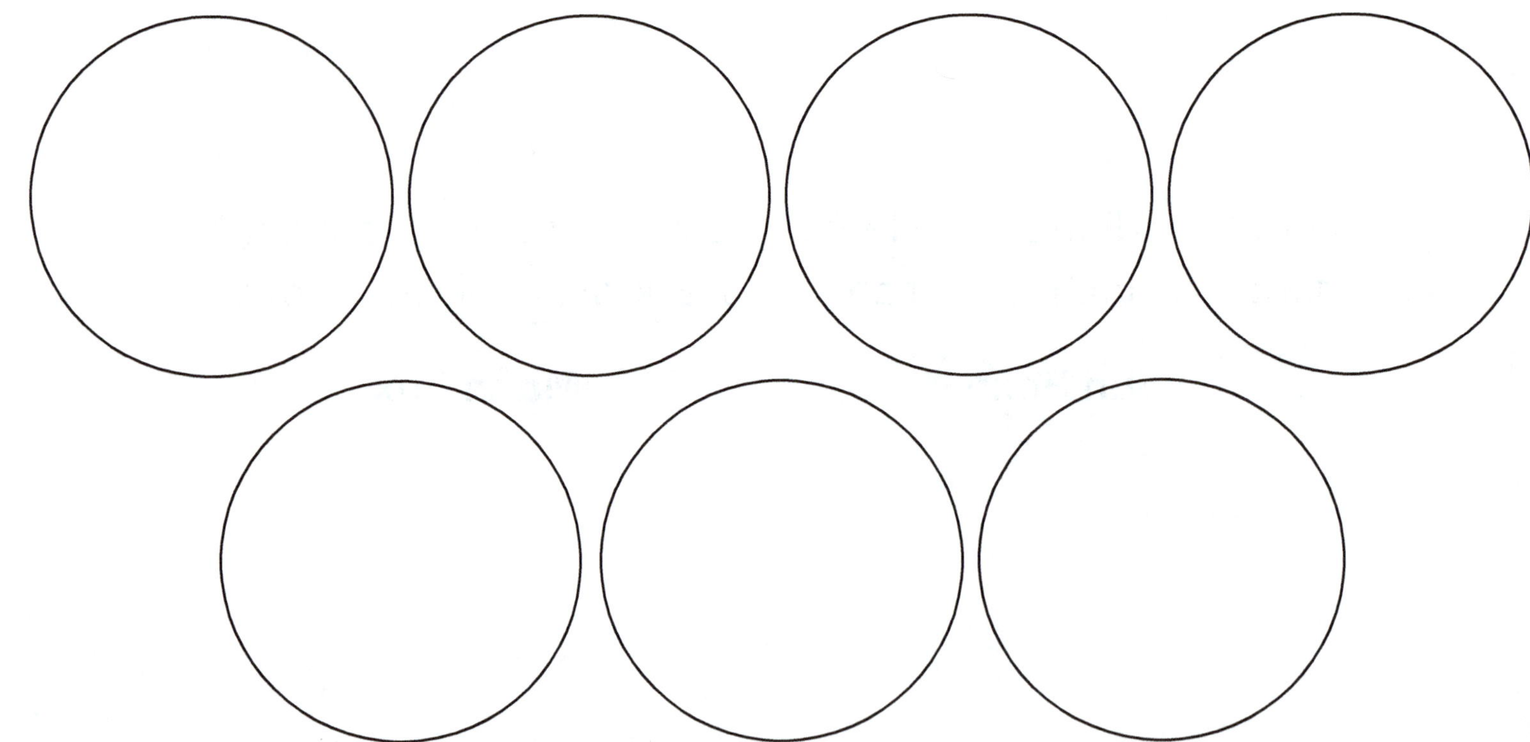

13 Living and Non-living Things

We Will Learn

- Living Things
- Non-living Things
- Natural Non-living Things
- Man-Made Non-living Things
- Uses of Non-living Things

Rack Your Brain

Naina and Rima were sitting in the park. Naina picked up a stone and threw it into the pond. Rima saw a butterfly and ran after to catch it, but failed. Why was Naina able to pick up the stone? Why could Rima not catch the butterfly?

Let us read the reasons for them in the chapter.

Living Things

The things which have life are called living things. Given below are the characteristics of living things.

1. Living things breathe.

2. Living things need food.

3. Living things produce young ones.

4. Living things grow.

5. Living things move.

6. Living things feel.

Non-living Things

The things which do not have life are called non-living things.

They do not need food to eat, air to breathe and water to drink. They cannot move, feel, grow or reproduce on their own.

Car

Telephone

Fan

Rock

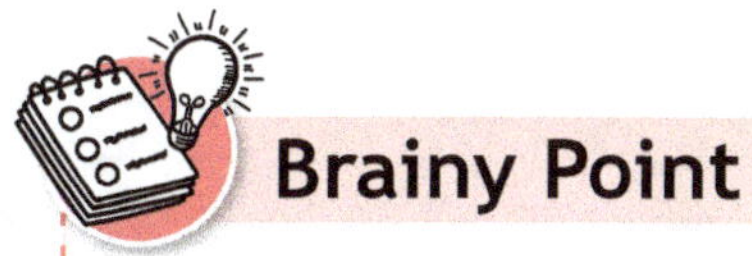

Brainy Point

All living things are natural. But non-living things can either be natural or made by man.

Non-living things are of two types:

Natural Non-living Things

The things which are gifted by Nature are called natural things. Water, wood, soil and rocks are some examples of natural non-living things.

Man-made Non-living Things

The things which are not found in Nature and made by man are called man-made non-living things.

Uses of Non-living Things

Non-living things are used by living things in many ways.

Living things use the light and heat given by the Sun.

- Water is used for drinking, bathing, cooking and washing clothes.
- Cars, utensils and many other things are made up of plastic, iron, steel, etc. All these non-living things are made by man.

Summary of the Chapter

- Everything around us is either a living or a non-living thing.
- All living things move, grow, eat and respond to the surroundings.
- All man-made things are non-living things.

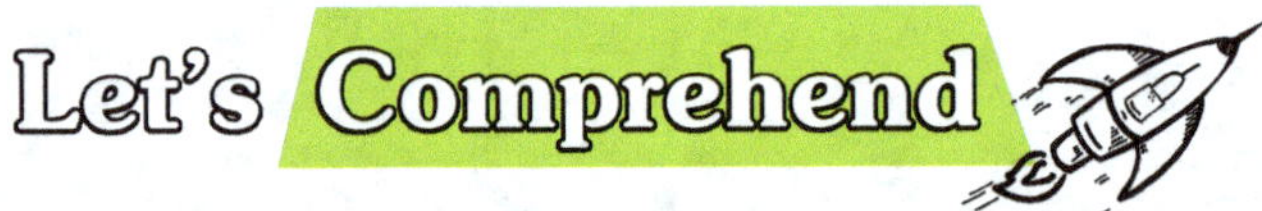

A. Answer the following questions :

1. What do you mean by living things ?

____

2. What do you mean by non-living things?

____

3. What are natural things?

____

4. What are man-made things?

____

B. Fill in the blanks :

1. Plants are _______________ things.

2. A rock is a ______________ thing.
3. Non-living things ______________ life.
4. The Moon is an example of a ______________ non-living thing.
5. The house is an example of a ______________ non-living thing.

C. Give two examples for each of the following :

1. Living things ______________ ______________
2. Non-living things ______________ ______________
3. Natural things ______________ ______________
4. Man-made things ______________ ______________

D. Encircle the correct pictures :

Get Hands-on Experience...

Take a round in the school compound. See different materials and group them into living and non-living things.

Hands-on Learning...

Collect and paste five living things and five non-living things in the space given below:

Living Things	Non-Living Things

14 Housing and Clothing

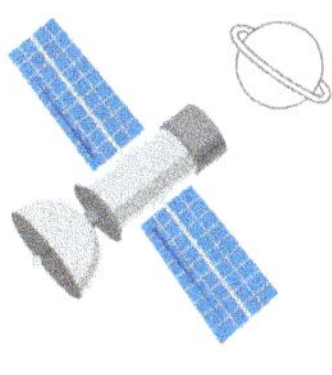

We Will Learn

- Need for a House
- Rooms in a House
- Need for Clothes
- Clothes in the Summer Season
- Clothes in the Winter Season
- Clothes in the Rainy Season

Rack Your Brain

Vinay lives in a very big house whereas his classmate Shiva lives in a small house. Vinay asks him about the rooms he has in his house. They conclude that Vinay has an extra room as compared to Shiva's house. Similarly, Mini saw many kinds of clothes in a shopping mall and was surprised why we need them.

Let's read all about the clothes and houses in the chapter.

Need for a House

Each of us needs a house to live in. A house protects us from heat, rain, strong winds and cold. It also protects us from thieves and wild animals.

Rooms in a House

A good, well-built house has many rooms. Each room is used for a different purpose.

Bedroom

We sleep and rest in a bedroom.

Drawing Room

We spend time with the members of our family and friends in the drawing room.

Dining Room

We take our meals together in the dining room.

Study

We study in the study. We should keep all books and our bag in the study.

Kitchen

Food is cooked in the kitchen.

Washroom

We take bath in a washroom.

We should keep our house neat, clean and airy. Things should be cleaned and kept at proper places to make the house tidy.

We should not throw garbage everywhere in the house. It should be thrown in the dustbin only.

Need for Clothes

We wear clothes to cover our bodies. Clothes protect us from heat, cold, rain, dust and dirt. We wear different types of clothes in different seasons.

Clothes in the Summer Season

We wear light cotton clothes in the summer season. Cotton clothes protect us from the heat of the Sun and keep us cool. Cotton clothes are made from cotton.

Clothes in the Winter Season

We wear woollen clothes in the winter season. Woollen clothes protect us from the cold wind and keep us warm. These clothes are made from wool.

Brainy Point

Cashmere wool is made from fibres we get from the Kashmiri (cashmere) goat.

Clothes in the Rainy Season

In the rainy season, we need an umbrella or a raincoat to keep us dry. We also wear gumboots to protect us from getting wet.

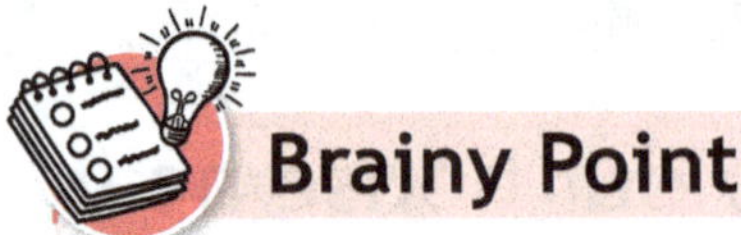

Brainy Point

We get wool from a sheep.

Brainy Point

The early man lived in a cave. He used to wear the bark of a tree and an animal skin to cover his body.

Some people who do special jobs such as doctor, lawyer, policeman, etc. wear specific clothes. These clothes are called uniforms. Kids also have to wear uniforms while going to school.

Summary of the Chapter

- Our house keeps us safe from the sun, heat, wild animals, enemies, etc.
- Our house has different rooms where we do different things.
- Clothes cover our bodies.
- We wear clothes according to weather.
- Some people who do special jobs wear uniforms.
- We wear cotton clothes in the summer season.
- We wear woollen clothes in the winter season.

Let's Comprehend

A. Answer the following questions :

1. Why does each of us need a house ?

____

2. Why do we need clothes ?

____

3. What types of clothes do we wear in the winter season?

____

B. Fill in the blanks :

1. A ______________ , ______________ house has many rooms.
2. ______________ is cooked in the kitchen.
3. ______________ clothes keep us warm.
4. Kids have to wear ______________ while going to school.
5. We must keep our house ______________ and ______________.

C. Give one word for each :

1. Clothes that keep us cool. ______________
2. Clothes that keep us warm. ______________
3. Room in a house where we eat meals. ______________
4. The thing used for protecting you from the rain. ______________

Get Hands-on Experience...

Your old clothes and toys can bring smiles to many poor children in orphanages. With the help of your parents, collect old clothes and toys that you no longer use. Also, collect the same from your friends and neighbours. Donate these old clothes and toys to the children living in orphanages.

Hands-on Learning...

Collect the pictures of various rooms in a house. Paste them in the given box:

Assessment Sheet - I

A. Answer the following questions briefly :

1. What do you mean by plant-eating animals?

   ____

2. What are dried fruits?

   ____

3. What is the difference between a creeper and a climber?

   ____

4. What are natural things?

   ____

5. Where does a woodpecker live?

   ____

B. Fill in the blanks :

1. Frogs and lizards eat ______________.
2. A rock is a ______________ thing.
3. ______________ are short and bushy plants.
4. ______________ are our green friends.
5. A ______________ does not have wings.

C. Name two things that you see :

1. in a park ______________ ______________
2. in the classroom ______________ ______________
3. on the road ______________ ______________

D. Match the following :

	Column A		Column B
1.	Bird that can fly	a.	Beetle
2.	Bird that cannot fly	b.	Moth
3.	Animal that lives on land	c.	Emu
4.	Animal that lives in water	d.	Elephant
5.	Insect that can fly	e.	Pigeon
6.	Insect that cannot fly	f.	Hippopotamus

E. Tick (✓) the correct answer :

1. This part is eaten as food ______________ .

   a. Stem ☐ b. Leaf ☐ c. Fruit ☐

2. The Moon is an example of a ______________ .

   a. living thing ☐

   b. man-made thing ☐

   c. natural thing ☐

3. It helps the plant to breathe:

   a. Water ☐ b. Air ☐ c. Sunlight ☐

4. The plants that grow along the ground are called ______________ .

   a. Climbers  b. Creepers  c. Shrubs 

5. We get coffee from ______________ .

   a. plants  b. factories  c. animals 

Assessment Sheet - II

A. Answer the following questions briefly :

1. Why does each of us need a house?

   ____

2. From where do we get water?

   ____

3. Why should we take proper rest?

   ____

4. Why do the stars appear so tiny to us?

   ____

5. What do you mean by wind?

   ____

B. Fill in the blanks :

1. ______________ is the main source of water.
2. Air occupies ______________ .
3. Do not leave your toys on the ______________ .
4. ______________ clothes keep us warm.
5. We must keep our house ______________ and ______________ .

C. Write (T) for a true statement and (F) for a false one :

1. We must run on stairs. ☐
2. Bamboo is the fastest growing plant on the Earth. ☐
3. We get vegetables and fruits from animals. ☐
4. We cross the road at the giraffe crossing. ☐
5. The blue whale is a marine mammal. ☐

D. Match the following :

Junk Food

Healthy Food

E. Tick (✓) the correct answer :

1. Turmeric, coriander and chilli are some examples of ________ .

   a. oilseeds ☐ b. pulses ☐ c. spices ☐

2. It is the main source of water.

   a. Handpump ☐ b. Rain ☐ c. Tap ☐

3. It gives us heat and light.

   a. Sun ☐ b. Moon ☐ c. Star ☐

4. These clothes keep us cool.

   a. Woollen ☐ b. Cotton ☐ c. Uniforms ☐

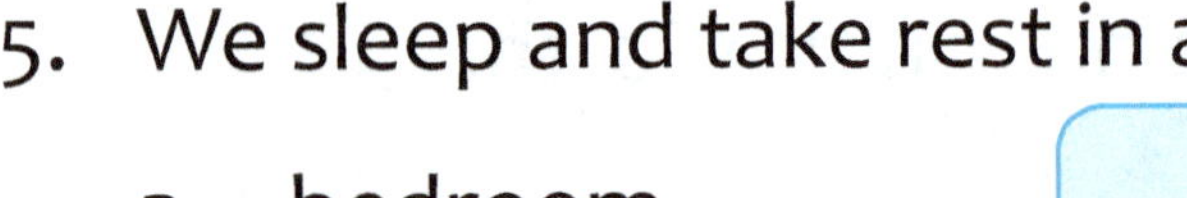

5. We sleep and take rest in a ________ .

   a. bedroom ☐

   b. drawing room ☐

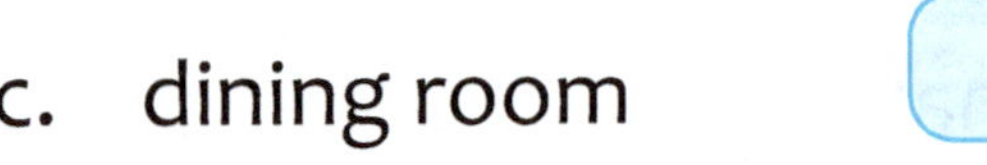

   c. dining room ☐

www.ingramcontent.com/pod-product-compliance
Lightning Source LLC
LaVergne TN
LVHW061939220826
846092LV00007B/1066

* 9 7 8 9 3 5 5 7 9 3 2 8 7 *